Gnats

Edited by Zeeshan Mahmud

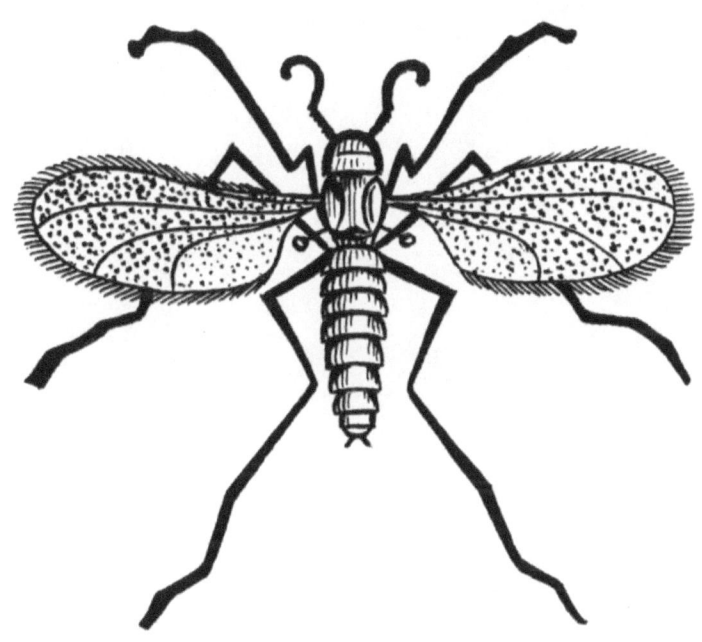

What are gnats?

Gnats are small, winged insects belonging to the order *Diptera*, which also includes flies and mosquitoes. While the term "gnat" does not refer to a single taxonomic group, it typically describes several species of tiny flying insects, including members of families like *Sciaridae* (fungus gnats), *Ceratopogonidae* (biting midges), and *Culicidae* (mosquitoes).

Physical Characteristics

- **Size**: Usually 1–3 mm long.
- **Appearance**: Slender bodies with long legs and delicate wings, often with vein patterns.
- **Color**: Varies from black to grayish or yellowish, depending on the species.

Habitat and Behavior

- **Habitats**: Moist environments, including forests, wetlands, soil, or decaying organic matter. Fungus gnats, for instance, thrive in damp soil and plant pots, while biting gnats prefer water-rich habitats.
- **Activity**: Most are active during the day, but some species are nocturnal.
- **Diet**:
    - Adult gnats may feed on nectar or other plant liquids.
    - Larvae typically feed on fungi, decaying organic matter, or plant roots.

Reproduction and Life Cycle

1. **Eggs**: Laid in moist or decaying organic matter.
2. **Larvae**: Emerge and feed on organic material or fungi, depending on the species.

3. **Pupae**: After a few molts, the larvae pupate in the soil or other substrates.
4. **Adults**: Live for a few days to weeks, focusing on reproduction.

Ecological Role

- **Decomposers**: Many gnats contribute to nutrient cycling by breaking down organic matter.
- **Pollinators**: Some species aid in pollination.
- **Food Source**: Serve as prey for larger insects, birds, and amphibians.

Nuisance and Control

- **Pests**: Fungus gnats can damage plant roots, while biting gnats (e.g., black flies) can be annoying or transmit diseases.
- **Control**:
    - Reduce moisture in indoor plants or soil.
    - Use sticky traps or biological agents like *Bacillus thuringiensis* (*Bt*).
    - Ensure proper drainage and avoid overwatering.

Gnats are a diverse group, with their biology and behavior varying widely across species.

Gnats and species

Gnats encompass thousands of species across several families, with the exact number being difficult to determine due to their diverse classifications and ongoing discoveries. Below are notable families and species of gnats, with descriptions highlighting their ecological roles or impact on humans:

Notable Gnat Families and Species

1. **Fungus Gnats (Family: *Sciaridae* and *Mycetophilidae*)**
   - **Species Count**: Over 2,000 described species globally.
   - **Notable Species**:
     - *Sciara militaris*: Known for mass synchronized larval migration, often forming "armyworm" trails.
     - *Bradysia impatiens*: A common pest in greenhouses, damaging roots of plants.
   - **Description**: Typically small (2–5 mm) with dark bodies and long legs. Larvae feed on fungi and decaying organic matter but may damage plant roots.
2. **Biting Midges (Family: *Ceratopogonidae*)**
   - **Species Count**: Approximately 5,000 species worldwide.
   - **Notable Species**:
     - *Culicoides imicola*: A vector for bluetongue virus in livestock.
     - *Leptoconops torrens*: A painful biting midge found in the U.S.
   - **Description**: Tiny gnats (1–3 mm) with piercing-sucking mouthparts. Adults are blood-feeders, particularly on livestock and humans.

3. **Black Flies (Family: *Simuliidae*)**
   - **Species Count**: Around 2,200 species identified.
   - **Notable Species**:
     - *Simulium damnosum*: Transmits river blindness (*Onchocerciasis*) in Africa.
     - *Simulium venustum*: A biting nuisance in North America.
   - **Description**: Robust-bodied gnats (2–5 mm) with humpbacked thoraxes. Larvae are aquatic and attach to submerged vegetation or rocks.
4. **Gall Gnats (Family: *Cecidomyiidae*)**
   - **Species Count**: Over 6,000 described species.
   - **Notable Species**:
     - *Mayetiola destructor* (Hessian fly): A major agricultural pest of wheat.
     - *Contarinia nasturtii* (Swede midge): Damages brassica crops like broccoli and cauliflower.
   - **Description**: Delicate-bodied gnats (1–4 mm). Some induce galls (abnormal plant growths) as larvae.
5. **Eye Gnats (Family: *Chloropidae*)**
   - **Species Count**: Estimated at 1,000 species.
   - **Notable Species**:
     - *Liohippelates collusor*: Common in tropical and subtropical areas, attracted to mucus and wounds.
   - **Description**: Small (1–2 mm), shiny gnats. Non-biting but cause irritation by feeding on bodily fluids.
6. **Sand Flies (Family: *Psychodidae*, Subfamily *Phlebotominae*)**
   - **Species Count**: Over 700 species in this subfamily.
   - **Notable Species**:
     - *Lutzomyia longipalpis*: A vector for leishmaniasis in South America.

- *Phlebotomus papatasi*: Transmits cutaneous leishmaniasis in the Middle East.
  - **Description**: Tiny (1.5–3.5 mm) with hairy bodies and wings, known for their painful bites.

Interesting Facts

- **Diversity**: Gnats are not monophyletic, meaning they do not form a single evolutionary lineage. Instead, they represent an informal grouping of small, delicate flies.
- **Behavioral Traits**: Some species form swarms during mating, creating visible "dancing" clouds.
- **Ecological Importance**: Fungus gnats aid in decomposition, and gall-inducing species influence plant morphology, often to their advantage.

Gnats' remarkable diversity makes them significant in ecology, agriculture, and public health.

Did 'Gnat' see that coming?

Gnats have been the subject of numerous scientific studies due to their ecological importance, agricultural impact, and role in vector-borne diseases. Below are key areas where gnats have been studied:

1. Ecology and Biodiversity

- **Role in Ecosystems**: Studies have highlighted the role of gnats in nutrient cycling and food webs. For example, fungus gnats (*Sciaridae*) contribute to organic matter decomposition by feeding on fungi and decaying vegetation.
- **Aquatic Larvae of Black Flies**: Research has focused on the role of black fly larvae (*Simuliidae*) in improving water quality by filtering suspended particles.
- **Biodiversity Research**: Studies of gall gnats (*Cecidomyiidae*) have shown their influence on plant evolution due to their ability to induce galls, leading to complex plant-insect interactions.

2. Agricultural Impact

- **Fungus Gnats in Greenhouses**:
    - *Bradysia species* have been studied extensively for their impact on crops like lettuce, tomatoes, and ornamental plants.
    - Research has explored biological controls, including *Bacillus thuringiensis* (*Bt*) and predatory nematodes, to manage infestations.

- **Gall Induction by Cecidomyiids**:
    - The Hessian fly (*Mayetiola destructor*) is one of the most studied agricultural pests. Studies focus on its interactions with wheat, including plant resistance mechanisms and pest control strategies.
- **Crop Damage from Swede Midge**:
    - *Contarinia nasturtii* has been studied for its economic impact on brassica crops. Research includes integrated pest management (IPM) strategies.

---

3. Disease Transmission

- **Vector of Human and Animal Diseases**:
    - Black flies (*Simuliidae*) have been studied extensively as vectors of *Onchocerca volvulus*, the parasitic worm that causes river blindness (onchocerciasis). Research has led to successful control programs, such as ivermectin distribution campaigns.
    - Biting midges (*Ceratopogonidae*, genus *Culicoides*) have been studied for their role in transmitting bluetongue virus and Schmallenberg virus, both of which affect livestock.
- **Sand Flies and Leishmaniasis**:
    - Species like *Phlebotomus papatasi* have been a focus of research for their role in spreading *Leishmania* protozoa. Studies explore sand fly biology, control methods, and vaccine development for leishmaniasis.

---

4. Behavior and Physiology

- **Mating Swarms**:
  - Swarming behavior in gnats, particularly biting midges and fungus gnats, has been studied to understand mechanisms of mate selection and population dynamics.
- **Larval Feeding Mechanisms**:
  - Research on fungus gnat larvae (*Bradysia species*) has elucidated how they digest organic matter and fungi, revealing potential methods for pest control.
- **Diapause in Black Flies**:
  - Studies have examined diapause (a dormancy period) in black fly larvae to understand how populations survive unfavorable environmental conditions.

5. Pest Management and Biological Control

- **Biological Agents**:
  - Studies on *Bacillus thuringiensis var. israelensis* (Bti) have demonstrated its effectiveness in controlling larvae of black flies and fungus gnats.
- **Predators and Parasitoids**:
  - Research has identified beneficial predators like rove beetles (*Dalotia coriaria*) and parasitoid wasps as effective against gnats in controlled environments.
- **Repellents and Attractants**:
  - Studies have evaluated essential oils, pheromones, and synthetic attractants for managing biting gnats and midges.

6. Climate Change and Distribution

- **Shifting Habitats**:
  - Research indicates that the distribution of gnats, especially biting midges (*Culicoides*), is influenced by climate change, with warming temperatures expanding their range and increasing disease transmission risks.
- **Phenology Studies**:
  - Studies on seasonal patterns in gnat populations help predict outbreaks of agricultural pests and disease vectors.

---

7. Molecular Studies

- **Genomics**:
  - Genome sequencing of species like the Hessian fly has provided insights into pest adaptation and resistance mechanisms in plants.
- **Microbial Symbiosis**:
  - Research on symbiotic bacteria in gnats (e.g., *Wolbachia*) has explored their role in reproduction and potential for biocontrol.

Which scientists were 'doing' gnats?

The study of gnats and small flying insects began with early naturalists and scientists who developed tools to observe tiny organisms. Here's an overview of notable contributions:

1. Robert Hooke (1635–1703)

- **Contribution**: Hooke was one of the first scientists to use a microscope to study small organisms.
- **Documentation**: In *Micrographia* (1665), Hooke detailed microscopic observations of insects, including their wings and body structures. While he did not focus exclusively on gnats, his work laid the foundation for observing small Diptera.

2. Antonie van Leeuwenhoek (1632–1723)

- **Contribution**: Van Leeuwenhoek, often regarded as the father of microbiology, used his handcrafted microscopes to study and describe tiny organisms.
- **Relevance to Gnats**: He observed and documented insect anatomy and behavior, including descriptions of winged insects similar to gnats.

3. Carl Linnaeus (1707–1778)

- **Contribution**: Linnaeus formalized the binomial nomenclature system in *Systema Naturae* (1758), which classified various species of gnats and their relatives within the order *Diptera*.
- **Notable Work**: He described early species of biting midges (*Ceratopogonidae*) and other small flies.

4. Johann Wilhelm Meigen (1764–1845)

- **Contribution**: Known as the father of Dipterology, Meigen made extensive studies of the order *Diptera*.
- **Relevance**: He described numerous gnat species and categorized their families systematically in his multi-volume work on flies.

5. Patrick Manson (1844–1922)

- **Contribution**: A pioneering figure in medical entomology, Manson linked biting flies, including gnats like black flies (*Simuliidae*), to the transmission of parasitic diseases.
- **Relevance**: His research laid the groundwork for understanding vector-borne diseases, including those spread by gnats.

Documented Studies and Records

- **Historical Cases**:
    - Early agricultural writings from the 18th century documented damage caused by fungus gnats and gall gnats, especially in Europe.

- Scientific journals in the 19th century reported outbreaks of biting midges and black flies, emphasizing their role as pests and disease vectors.

While Hooke and van Leeuwenhoek pioneered the microscopic study of tiny insects, Linnaeus and Meigen formalized the classification of gnats, and later scientists like Manson explored their medical and ecological significance. These contributions collectively advanced our understanding of gnats.

Gnatomy

Gnats, as small Dipteran insects, share a typical anatomy with other flies, modified for their size and ecological roles. Below is a detailed breakdown of gnat anatomy using scientific terms:

External Anatomy

1. Head

- **Antennae**:
    - Multisegmented, filiform or moniliform (bead-like), with sensilla that aid in detecting pheromones, odors, and humidity.
    - Longer and more elaborate in males for detecting female signals in mating species like *Ceratopogonidae*.
- **Eyes**:
    - **Compound Eyes**: Large and hemispherical, composed of ommatidia, specialized for motion detection.
    - **Ocelli**: Simple eyes (usually three) located dorsally between the compound eyes, aiding in light perception.
- **Mouthparts**:
    - **Non-Biting Gnats** (*Sciaridae, Cecidomyiidae*):
        - Labium and maxillae form a proboscis adapted for feeding on plant sap or fungi.
    - **Biting Gnats** (*Ceratopogonidae, Simuliidae*):
        - Labrum and mandibles are modified into piercing structures for blood-feeding (haematophagy).

- Hypopharynx injects anticoagulants while the labium forms a sheath around other parts.

2. Thorax

- **Prothorax, Mesothorax, Metathorax:**
    - Segments fused to form a rigid structure for muscle attachment.
    - Mesothorax enlarged to house flight muscles, as gnats are strong fliers despite their small size.
- **Legs:**
    - Long, slender legs with **coxae, trochanters, femurs, tibiae,** and **tarsi**.
    - Tarsi terminate in pulvilli and claws for gripping smooth surfaces.
- **Wings:**
    - Single pair of membranous wings (forewings):
        - Venation reduced, with key veins like **costa**, **subcosta**, and **radius** supporting wing structure.
        - Some species (*Psychodidae*) have hairy wings, aiding in flight stability.
    - Hindwings reduced to **halteres**:
        - Knob-like structures that function as gyroscopes for balance during flight.

3. Abdomen

- Composed of 10 segments:
    - Visible segments often reduced to 8 in adults, with the terminal segments forming the genitalia.
- **Genitalia:**

- Male: **Claspers** (gonostyli) used for mating; **aedeagus** for sperm transfer.
- Female: Ovipositor may be elongated for egg deposition into substrates like soil or plant tissues.
- **Spiracles**: Paired openings on lateral surfaces for respiration.
- **Cerci**: Short, sensory appendages at the tip of the abdomen in some species.

Internal Anatomy

1. Nervous System

- Centralized brain (supraesophageal ganglion) in the head, connected to a ventral nerve cord.
- Subesophageal ganglion controls mouthparts.
- Thoracic ganglia coordinate wing and leg movements.

2. Digestive System

- **Foregut**:
    - Mouth leads to pharynx and esophagus, which pass food to the crop for temporary storage.
    - Proventriculus regulates food passage to the midgut.
- **Midgut**:
    - Principal site for digestion and absorption. Contains specialized enzymes for breaking down sugars or blood proteins (in biting species).
- **Hindgut**:
    - Malpighian tubules excrete nitrogenous wastes into the gut.
    - Rectum reabsorbs water before excretion via the anus.

3. Circulatory System

- Open system with hemolymph (insect "blood") circulating through a dorsal vessel.
- Heart (posterior) and aorta (anterior) pump hemolymph throughout the body, delivering nutrients and removing waste.

---

4. Respiratory System

- **Tracheal System**:
    - Network of chitin-lined tubes delivering oxygen directly to tissues.
    - Spiracles located along the thorax and abdomen allow gas exchange.
- **Air Sacs**: Expandable structures in some species to enhance ventilation during activity.

---

5. Reproductive System

- **Male**:
    - Testes produce sperm, stored in seminal vesicles.
    - Accessory glands produce seminal fluid.
- **Female**:
    - Paired ovaries produce eggs, passed into the oviduct.
    - Spermatheca stores sperm post-mating for fertilization.

Specialized Adaptations

- **Biting Species**:

- Mouthparts adapted for piercing skin and extracting blood.
- Salivary glands produce anticoagulants to facilitate feeding.
- **Fungus Gnats**:
    - Larvae have specialized mandibles for feeding on mycelium in moist environments.

This level of detail demonstrates the intricate morphology and functionality of gnats, enabling them to thrive in various ecological niches.

'Gnats!' in various foreign languages

1. **French**: moucheron
2. **Spanish**: mosquito pequeño
3. **German**: Mücke
4. **Italian**: moscerino
5. **Portuguese**: mosquitinho
6. **Russian**: мошка (*moshka*)
7. **Chinese (Simplified)**: 蚋 (*ruì*)
8. **Japanese**: ブユ (*buyu*)
9. **Korean**: 각다귀 (*gakdagwi*)
10. **Arabic**: بعوضة صغيرة (*baʿūḍah ṣaghīrah*)
11. **Hindi**: छोटी मक्खी (*chhoti makkhī*)
12. **Bengali**: ক্ষুদ্র মশা (*kshudra mosha*)
13. **Turkish**: sivrisinek
14. **Greek**: σκνίπα (*sknípa*)
15. **Dutch**: mugje
16. **Swedish**: knott
17. **Norwegian**: knott
18. **Danish**: knort
19. **Finnish**: mäkärä
20. **Polish**: muszka
21. **Hungarian**: szúnyog
22. **Czech**: komárek
23. **Thai**: แมลงหวี่ (*malɛɛng wî*)

Gnat control

Call the g-unit! Or rather *unat*... you heard. Gnat control methods target both the larvae and adult stages, employing environmental, biological, chemical, and physical strategies. Here's a breakdown:

1. Environmental Control

- **Reduce Moisture**:
    - Eliminate standing water and damp soil, as gnats lay eggs in moist conditions.
    - Improve drainage in gardens or potted plants.
- **Proper Sanitation**:
    - Dispose of decaying organic matter (e.g., overripe fruit, compost) to remove breeding sites.
    - Clean garbage bins and food preparation areas.

2. Biological Control

- **Beneficial Organisms**:
    - **Nematodes** (*Steinernema feltiae*): Parasitic nematodes can target gnat larvae in the soil.
    - **Predatory Insects**: Lacewings and predatory mites feed on gnat larvae.
- **Microbial Control**:
    - **Bacillus thuringiensis var. israelensis (BTI)**: A naturally occurring bacterium that kills gnat larvae when applied to standing water or soil.

3. Chemical Control

- **Insecticides**:
    - Pyrethroid-based sprays for adult gnats.
    - Soil drenches with larvicides like imidacloprid for larvae.
- **Growth Regulators**:
    - Inhibitors like methoprene disrupt larval development.

Note: Use chemical methods sparingly and according to label instructions to minimize environmental impact.

---

4. Physical Control

- **Traps**:
    - **Sticky Traps**: Yellow adhesive traps attract and catch adult gnats.
    - **Homemade Vinegar Traps**: Bowls of vinegar mixed with a drop of dish soap attract and drown gnats.
- **Screens and Barriers**:
    - Install fine mesh screens to prevent gnats from entering indoor spaces.

---

5. Cultural Practices

- **Plant Management**:
    - Allow soil to dry out between waterings to deter larvae.
    - Avoid overwatering plants.
- **Food Storage**:
    - Keep fruits and vegetables stored in sealed containers.

Integrated Pest Management (IPM)

Combining these methods ensures effective and sustainable gnat control while minimizing harm to beneficial organisms and the environment.

Etymology of the word 'gnat'

The word *gnat* has an ancient origin rooted in Germanic languages. Here's a breakdown of its etymology:

1. **Old English**:
   - The word *gnæt* (pronounced similar to "gnat") referred to a small flying insect, particularly biting insects.
   - Related to the Old Norse word *gnattr*, which also described small insects.
2. **Proto-Germanic**:
   - Derived from *gnattaz*, meaning "gnawer" or "biter," reflecting the biting behavior of some species.
3. **Proto-Indo-European (PIE)**:
   - Possibly linked to the PIE root *gnē-* or *gnō-*, meaning "to know" or "to perceive," metaphorically connected to the sensory irritation caused by gnats.
4. **Modern Usage**:
   - The word has remained relatively unchanged in English since its Old English form, with cognates in other Germanic languages like Dutch (*gnat*) and Low German (*gnat*), though modern German uses *Mücke* for similar insects.

The silent "g" in *gnat* is a holdover from Old English and reflects early Germanic pronunciation conventions.

Proverbial usage of the word 'gnat'

Here are some proverbs and idiomatic expressions involving gnats:

1. **"Strain at a gnat and swallow a camel"**
   - Origin: Biblical (Matthew 23:24, KJV).
   - Meaning: Focus on insignificant details while ignoring major issues.
2. **"Gnats follow the sweet; wasps the sour"**
   - Meaning: Small irritations or troubles are drawn to things of value or sweetness, just as gnats are attracted to sweetness.
3. **"Small as a gnat's eyebrow"**
   - Origin: Often humorous or colloquial.
   - Meaning: Something extremely tiny or insignificant.
4. **"The gnat in the ear makes more noise than the lion in the forest"**
   - Meaning: Small, persistent annoyances can feel more disruptive than large, distant threats.
5. **"The gnat perishes in the honey"**
   - Origin: Derived from older European sayings.
   - Meaning: Excessive indulgence in pleasure can lead to harm or downfall.

These proverbs reflect how gnats symbolize small, often irritating things in life.

The phrase "attention span of a gnat" is a metaphor often used to describe extremely short attention spans. However, there is no scientific evidence specifically assessing the "attention span" of gnats. Insects like gnats rely on instinct and sensory inputs rather than deliberate focus or sustained attention, so the term is more figurative than biological.

Scientific Insights on Gnat Behavior

1. **Sensory Responsiveness:**
   - Gnats have highly sensitive antennae and eyes, enabling rapid reactions to stimuli like movement, light, or chemical signals.
   - Their responses are reflexive rather than indicative of attention or cognitive focus.
2. **Flight Behavior:**
   - Gnats exhibit hovering and erratic flight patterns, often interpreted as "distraction." This behavior is primarily driven by survival needs like finding mates or locating food sources.

Longevity of Gnats

The lifespan of a gnat depends on its species and environmental conditions:

1. **Adult Stage**:
    - Non-biting gnats (*e.g., Fungus Gnats*): Live about **1–2 weeks**.
    - Biting gnats (*e.g., Black Flies*): Live about **2–3 weeks**, often depending on blood meals for reproduction.
2. **Life Cycle**:
    - **Egg Stage**: 2–3 days.
    - **Larval Stage**: 1–2 weeks (longer in cool environments).
    - **Pupal Stage**: 3–6 days.
3. **Environmental Impact**:
    - Temperature, humidity, and food availability significantly affect the gnat's lifecycle and longevity.

While gnats are short-lived, their survival strategies ensure their persistence across generations.

Lifecycle of a gnat

The life cycle of gnats typically follows a four-stage process: **egg**, **larva**, **pupa**, and **adult**. The duration of each stage can vary depending on the species and environmental conditions, but here's a general overview:

1. Egg Stage

- **Duration**: A few days to a week, depending on species and environmental conditions (moisture and temperature are key factors).
- **Location**: Eggs are usually laid in moist environments, such as on the surface of water, decaying organic matter, or soil.
- **Egg Characteristics**:
    - Most gnat species lay eggs in clusters or individually. Some species, like fungus gnats, lay eggs in moist soil where fungal growth occurs, providing a food source for larvae.

---

2. Larval Stage

- **Duration**: Typically 1–2 weeks, but can be longer in cooler conditions.
- **Appearance**:
    - Gnats in their larval stage are worm-like and usually lack well-developed eyes or wings.
    - Fungus gnat larvae are translucent and often have a distinct black head.
- **Feeding**:

- Larvae are primarily herbivores or detritivores. Some species (like biting gnats) feed on organic matter, while others (like black flies) feed on blood or plant juices.
- **Habitat**:
  - Larvae live in moist, organic-rich environments, such as soil, decomposing vegetation, or water.

---

3. Pupal Stage

- **Duration**: Usually lasts 2–6 days, but can vary based on temperature and species.
- **Appearance**:
  - Pupae are the transitional form between larvae and adults. They are often immobile and encased in a protective casing called a **pupal exuvia**.
- **Behavior**:
  - Pupae typically stay in the same location, suspended in water or within the substrate.
  - During this stage, internal development of the adult body, including wings and reproductive organs, occurs.

---

4. Adult Stage

- **Duration**: Typically 1–3 weeks, though it varies by species and environmental conditions.
- **Appearance**:
  - Adult gnats are small, flying insects with delicate, often transparent wings.

        - Males and females typically have differing physical features, especially in species with sexual dimorphism.
- **Reproduction**:
    - Adults emerge from the pupal casing and begin mating shortly after emerging.
    - Female gnats typically require a blood meal to reproduce (in biting species), while non-biting species feed on nectar or other organic matter.
- **Behavior**:
    - Adult gnats spend much of their time flying in search of food, mates, or breeding sites. They can be attracted to light or specific chemicals, such as those produced by plants or animals.

General Lifespan of Gnats

- The total life cycle from egg to adult can range from **2 weeks to a month**, depending on environmental factors. In favorable conditions, some species can complete a cycle in as little as **10 days**.

This brief life cycle helps gnats proliferate rapidly in environments where conditions are optimal for their growth and reproduction.

No-see-ums?

No-see-ums (also known as **biting midges**, **sand flies**, or **punkies**) are tiny, biting insects in the family **Ceratopogonidae**, closely related to gnats. They are notorious for being difficult to see due to their small size, which is why they're called "no-see-ums."

Characteristics of No-See-Ums:

1. **Size**:
    - Typically **1–3 mm** long, which makes them nearly invisible to the naked eye, especially in flight.
    - Their small size contributes to their nickname, as they can bite without being easily noticed.
2. **Appearance**:
    - No-see-ums are tiny, with a slender body and wings that may be lightly covered in scales.
    - They are often grayish or translucent.
3. **Habitat**:
    - Found in damp or moist environments, particularly near water, such as marshes, beaches, and wetlands.
    - They are especially active at dawn and dusk.
4. **Feeding Behavior**:
    - Both males and females feed on nectar, but **females** are the ones that bite, as they require blood to produce eggs.
    - Their bites can be painful and cause itching, swelling, and irritation. Some species may transmit diseases in certain regions, such as **blue tongue virus** in livestock.
5. **Distribution**:

- No-see-ums are found worldwide, especially in coastal areas, wetlands, and places with high humidity.

Life Cycle:

Their life cycle follows similar stages to gnats, with eggs laid in moist or decaying organic material. The larvae develop in soil, leaf litter, or water, and they mature into tiny adults.

Control Methods:

- Reducing standing water and eliminating breeding sites.
- Using insect repellents containing DEET or picaridin.
- Installing screens to keep them out of homes and outdoor spaces.
- Wearing protective clothing during peak activity hours.

Despite their small size, no-see-ums can be quite bothersome due to their itchy bites and ability to appear unnoticed.

Size

Gnats are typically very small insects, with varying sizes depending on the species. Here's a general breakdown:

1. **Common Non-Biting Gnats**:
   - **Size**: Around **1–3 mm** in length. These are the types often seen buzzing around lights or in damp environments.
2. **Biting Gnats** (e.g., *Ceratopogonidae* family):
   - **Size**: Generally **1–5 mm**, although some species may appear slightly larger when feeding or flying in swarms.
3. **Fungus Gnats**:
   - **Size**: Usually about **2–4 mm** in length, with long, delicate wings that make them appear slightly larger.

While they are all quite small, their size can vary somewhat, with the largest species reaching up to **6 mm** or slightly more. Despite their tiny size, their presence can be quite noticeable, especially in swarms.

Find some gnats fo' me

Gnats are highly diverse and are found in a wide range of habitats across the world. Their diversity spans over **1,000 species**, and they can thrive in nearly every environment, from tropical regions to temperate climates. Here's an overview of their diversity and global distribution:

1. Habitat Diversity

Gnats are adapted to a variety of ecosystems, including:

- **Wetlands and Marshes**: Many species, such as biting midges (*Ceratopogonidae*), thrive in wet, marshy areas with abundant organic matter.
- **Forests and Grasslands**: Gnats are commonly found in forests, where they can live in the moist soil and decaying organic material. Some species are attracted to certain types of plants or fungi in these environments.
- **Urban Areas**: Some species, like **fungus gnats**, are prevalent in greenhouses, gardens, or homes with potted plants where moisture and decaying organic matter are present.
- **Coastal Regions**: No-see-ums (biting midges) are particularly common in coastal areas, including beaches, tidal marshes, and mangrove swamps.
- **Tropical and Subtropical Regions**: Gnats are widespread in tropical regions, where high humidity and consistent warmth create ideal conditions for their breeding and survival.
- **Temperate Zones**: Gnats can also be found in temperate climates, especially in areas with high moisture, like lakes, rivers, or damp woods.

2. Global Distribution

Gnats are found all over the world, though their species diversity and abundance vary by region:

- **North America**:
    - Various species, like **fungus gnats** and **black flies**, are common in wetland areas and can be found across the continent, particularly in the northeastern United States and Canada.
    - In coastal areas, **no-see-ums** are prevalent in regions like the Gulf Coast and Pacific Northwest.
- **Europe**:
    - Europe has a rich diversity of gnat species, especially in moist environments. Species like the **midge** are widespread.
    - The **biting midge** (*Ceratopogonidae*) is especially common in countries like the UK, where they can be a nuisance during the warmer months.
- **Africa**:
    - Gnats are abundant in tropical and subtropical climates, especially near freshwater bodies.
    - Some species, like the **black fly**, are vectors for diseases such as **onchocerciasis** (river blindness) in parts of sub-Saharan Africa.
- **Asia**:
    - In countries like India and Southeast Asia, gnats are common in marshes, rice paddies, and tropical forests. Species diversity is high in these humid regions.
    - **No-see-ums** and other biting midges are found in coastal areas.
- **Australia**:
    - Gnats, especially biting species like the **sandfly**, are found in coastal and tropical regions, often in mangrove swamps or around beaches.

        - Species such as **Culicoides** are known to cause irritation in humans and animals.
- **South America**:
    - South America's tropical and wetland regions are home to a variety of gnat species, including both biting and non-biting types.
    - Countries like Brazil and Colombia experience issues with gnats in riverine and forested areas.

---

3. Key Species and Their Global Distribution

- **Fungus Gnats** (*Sciaridae*): Found worldwide, particularly in areas with decaying organic matter like compost or damp environments.
- **Biting Midges** (*Ceratopogonidae*): Widespread, with significant populations in tropical and temperate regions. Notably found in coastal and marshy areas.
- **Black Flies** (*Simuliidae*): Common in regions with fast-moving streams and rivers. They are more abundant in temperate climates but can be found worldwide.
- **No-See-Ums** (*Ceratopogonidae*): Found in tropical and coastal areas worldwide, especially in regions with marshes and tidal flats.

---

4. Environmental Impact

Gnats are an important part of ecosystems, especially in wetland and forest environments, where their larvae decompose organic matter and help recycle nutrients. However, some species, like biting midges, can be pests, transmitting diseases to both humans and animals. Their presence often correlates with humid and moist environments, making them a regular feature in ecosystems around the world.

Bizarro!

Here are some **true and interesting anecdotes** about gnats:

1. Gnats and the Spread of Disease

- Certain types of gnats, especially **biting midges** (*Ceratopogonidae*), are known vectors of diseases. One of the most notorious is the **blue tongue virus**, which affects livestock like sheep and cattle. In sub-Saharan Africa, **black flies** (*Simuliidae*), a type of gnat, transmit **onchocerciasis** (river blindness) to humans. It's a serious condition that has affected millions of people, particularly in rural regions near fast-flowing rivers.

2. Gnats as Super Pollinators

- Despite their small size, gnats play a significant role in pollination. **Fungus gnats**, which are attracted to decaying organic matter, can inadvertently transport pollen between plants. Interestingly, some plants that thrive in dark, damp environments rely on gnats and similar insects for pollination, including certain orchids. The plant *Lindernia* even produces a smell designed to attract fungus gnats, ensuring the pollination of its tiny flowers.

3. The Gnat's Unseen Defense

- While gnats are often seen as a nuisance to humans, they have fascinating **defensive** behaviors. Some species of gnats, such as **black flies**, release an irritating saliva when they bite, which can cause prolonged itching. The saliva contains anticoagulants, which prevent the blood from clotting while the gnat feeds. This mechanism not only helps the gnat feed but also serves as a defense against other predators that might try to consume them.

4. Fungus Gnats and Climate Change

- Fungus gnats have been found to be particularly **sensitive to climate change**. Studies show that as temperatures rise, their **reproductive cycles** may become faster, increasing their populations in urban and agricultural areas. This has led to an increase in the frequency of their presence in greenhouses, where they can harm plants by feeding on their roots.

5. Anecdote from the Antarctic

- Despite being associated with warmth and humidity, **gnats** have even been spotted in extreme environments like the **Antarctic**. Researchers have discovered species of midges, such as the **Antarctic midge** (*Belgica antarctica*), that live in the Antarctic's cold and dry conditions. This species is one of the only insects to inhabit the continent and has adapted to survive in freezing temperatures by slowing down its metabolism.

6. Biting Midges and "Swarm" Behavior

- Biting midges are known for their **swarming behavior**, where thousands of gnats will converge on an area to mate. This behavior often creates the illusion of a cloud of tiny insects. The **swarm** behavior is not only a mating ritual but also helps these gnats avoid predators. They gather in large numbers to overwhelm potential threats, making it difficult for birds or other insects to single out an individual gnat.

7. The "Gnat Dance"

- In many cultures, people have observed what seems like a "**dance**" of gnats swarming in the air. This behavior is actually a **mating ritual**. Males will fly in tight circles or create erratic patterns in the air to attract females. Sometimes, these displays can appear like an intricate performance, but they're simply the gnats' method of finding a mate.

These anecdotes highlight how gnats, despite their tiny size, have fascinating behaviors, ecological roles, and surprising adaptability that make them an interesting subject of study.

Gnats symbolism

Gnats appear in various mythologies, although they are not often central characters in ancient myths. However, there are some notable references to gnats or related insects in **ancient Egyptian mythology** and other cultures:

1. Ancient Egypt:

In ancient Egypt, gnats were sometimes associated with **pestilence** and **divine retribution**. They appear in one of the **Ten Plagues of Egypt**, a biblical account found in the Book of Exodus, where gnats (or "lice" in some versions) were sent by God to punish the Egyptians for not freeing the Israelites. The exact identity of the insect is debated, but gnats or lice are commonly considered to be the pestilence mentioned:

- **The Plague of Gnats**: One of the plagues involves the transformation of dust into gnats that would infest humans and animals, causing great discomfort. This is seen as a sign of divine wrath. The biblical passage is often interpreted as an indication of the power of the Hebrew God in bringing nature under control as a sign of punishment for the Pharaoh's refusal to let the Israelites go.

2. Other Cultures:

Gnats have appeared in different mythologies, often representing **insignificance**, **annoyance**, or **pestilence**. Their tiny size and large numbers can symbolize things that are bothersome yet too small to be significant. Some cultures use gnats as metaphors for fleeting or troublesome elements of life.

- **Ancient Greece and Rome**: In Greek mythology, gnats were sometimes viewed as **symbols of insignificance** or

pestilence, similar to the biblical plague. They weren't given a specific mythological role but were often associated with irritation or something that could be endured in a world of larger, more dramatic mythological figures.
- **Africa and Folklore**: Some African myths treat gnats as tricky creatures associated with **deceit** or small distractions in the natural world. While gnats themselves aren't central to these stories, their presence is often linked with the **frustrations** or **harms** brought about by seemingly insignificant things.

3. Symbolic Role in Literature and Folklore:

Gnats and similar small pests are also used symbolically in various **literary** and **folkloric traditions**. They are often used as metaphors for **insignificant annoyances** or **small evils**, often highlighting human vulnerability to forces that seem tiny but can cause significant discomfort.

www.ingramcontent.com/pod-product-compliance
Lightning Source LLC
Chambersburg PA
CBHW031555210526
45464CB00003B/1302